From
Presentations Kill?
To
Presentation Skill!

*7 Easy Steps for
Time-Challenged
Glossophobia Sufferers*

By

David J. Hobson

© Copyright 2005 David Hobson
All rights reserved. No part of this publication may be reproduced, stored in a retrieval system, or transmitted, in any form or by any means, electronic, mechanical, photocopying, recording, or otherwise, without the written prior permission of the author.

Note for Librarians: A cataloguing record for this book is available from Library and Archives Canada at www.collectionscanada.ca/amicus/index-e.html
ISBN 1-4120-7203-4

Printed in Victoria, BC, Canada. Printed on paper with minimum 30% recycled fibre. Trafford's print shop runs on "green energy" from solar, wind and other environmentally-friendly power sources.

TRAFFORD
PUBLISHING

Offices in Canada, USA, Ireland and UK
This book was published *on-demand* in cooperation with Trafford Publishing. On-demand publishing is a unique process and service of making a book available for retail sale to the public taking advantage of on-demand manufacturing and Internet marketing. On-demand publishing includes promotions, retail sales, manufacturing, order fulfilment, accounting and collecting royalties on behalf of the author.

Book sales for North America and international:
Trafford Publishing, 6E–2333 Government St.,
Victoria, BC v8t 4p4 CANADA
phone 250 383 6864 (toll-free 1 888 232 4444)
fax 250 383 6804; email to orders@trafford.com
Book sales in Europe:
Trafford Publishing (uk) Limited, 9 Park End Street, 2nd Floor
Oxford, UK OXI IHH UNITED KINGDOM
phone 44 (0)1865 722 113 (local rate 0845 230 9601)
facsimile 44 (0)1865 722 868; info.uk@trafford.com
Order online at:
trafford.com/05-2098
10 9 8 7 6 5 4 3

Contents

Introduction *1*
The Quest *2*
The Consultant *3*
The Coach *12*
Step 1: Research *25*
Step 2: Planning *33*
Step 3: Visual Aids *49*
Step 4: Techniques *57*
Step 5: Practice *77*
Step 6: Interaction *80*
Step 7: Breathe! *90*
Controlling nervousness *92*
The Review *97*
The Response *99*
Venue Check List *101*
Acknowledgements *105*
About The Author *107*

Introduction

Even in today's technological age, conferences, seminars, workshops and meetings will still require the personal touch of a presentation. Yet many people would still do just about anything to avoid making a presentation. They let fear get in the way of success

Another stumbling block is lack of time. The type of person who needs 30 hours every day just to complete their workload does not have time to wade through a How-To book the size of *War and Peace* to learn how to make effective, positive presentations. That is why this book is written with you, the busy person, in mind!

I hope you enjoy using this book and will also enjoy giving presentations as a result.

David Hobson

Frustration was what drove the young man to make an appointment with the Business Consultant.

The young budding entrepreneur had launched his own business venture and had been working hard at trying to make it, and himself, a success.

After many months of determination, marketing and working 24/7, a friend advised him to seek out the services of a professional; someone who could guide and coach him in the ways to achieve his desired results. Someone who had been there, done that and got the T-shirt. Someone who could be—a Mentor.

Frustrated beyond belief by his inability to achieve the success he thought he was capable of, the young man took his friend's advice and went looking for such a person through his contacts within the Chamber of Commerce. The name of one particular Business Consultant seemed to be at the top of many people's lists and so the young man took the plunge and made the appointment.

Looking around the Consultant's reception area the young man noticed the motivational posters decorating the walls. Attitude. Believe and Succeed. Commitment. Character. Courage. Dare to Soar. Vision. Words of wisdom backed by simple, yet inspirational pictures. Although these posters made him feel hopeful, the young man was still apprehensive as he waited for the Consultant.

Smiling broadly and extending a hand of welcome, the Business Consultant strode into the reception area and greeted the young man; he invited him into the inner sanctum of the office.

"So", said the Consultant, "tell me about your entrepreneurial efforts to date."

"Well, since I launched my business a year ago, I have been trying very hard to be successful. I have used all the recommended methods of sales and marketing.

"I drew up a business plan and projected my necessary outlays against anticipated income and projected a break-even point within 3 years. I watched other people in the same field and knew that with my skills and qualifications, plus my faith in the business, I could achieve the same results or better."

"Well", said the Consultant, "it certainly seems like you are on the right track. Making a plan is a good idea. It acts like a road map and helps you get to your destination. Observing the activities of your competitors is essential if you want to survive in business.

"Timing, too, plays a vital role in the business planning side of things; how are your administrative skills?"

The young man's eyes shone. "Oh I'm a very disciplined administrator", he said, "especially when it comes to time management. I carefully plan out each day in advance and I work out short, medium and long term schedules for achieving my goals."

"Goals are good. Tell me about some specific ones", the Consultant encouraged.

"My first goal was to establish a network link and I have achieved this through joining the local Chamber of Commerce as well as the Business Information Centre and the Better Business Bureau. I also have contacts with the Rotary Club and of course there is my Church. I regularly pass out my business card through all of these channels. But…" and here his face fell, "I do not seem to generate enough business."

The Consultant reflected on this for a few minutes and finally said softly, "It is, of course, very important to establish a network of contacts and to be constantly on the lookout for ways to use these contacts to promote your business.

"However, as you have found, this is not necessarily the best method. Now, tell me about how you have tried to market your services."

The initial nervousness of the young entrepreneur was now beginning to slip away as he sensed that his planning to date had been 'by the book'.

"That was my next goal", he said. "I designed a series of marketing leaflets in conjunction with an image designer, which have been circulated to a number of potential supportive outlets. I have a website up and running and I am investigating associate links with other complementary businesses."

"You *have* been busy", said the Consultant, smiling. "I am particularly glad to hear that you employed the services of an image designer to plan out your marketing materials. It is very important to create the right image in all your corporate brochures, web pages, letters and in fact, everything that leaves your office. Consistency and clarity reflect professionalism. So what is your 30 second, anti-boredom sound bite?"

"My what?"

"Your 30 second, anti-boredom sound bite. Sometimes known as an audio logo or an elevator speech."

"I'm sorry, I don't know what you mean." The young man's growing confidence now took a slight downward turn.

"If we met for the first time and I asked you what you did, you would have approximately 30 seconds to sell your business to me before I became bored. This sound bite is often called an audio logo. It is short and snappy and really reflects you and your business."

The young man looked slightly confused. "Let me give you an example" said the Consultant. "I could describe my consultancy business as 'Providing help, guidance and counselling on business oriented issues to established and aspiring entrepreneurs' or, more succinctly, I could say, 'I am a Business Mentor'."

"I see" reflected the young man. "I had not really thought of presenting myself that way. I can see that it is better than just saying, 'Here is my card'." The Consultant nodded. "If you are agreeable, I will come and talk to you again, when I have worked something out."

"Certainly; I will be happy to work with you as long as you need me. I like to establish ongoing relationships with my clients so that we can develop success and maintain it constantly. This is just the same way, hopefully, that you are conducting your own business.

"We have talked about your marketing materials and you have some homework to do, so now let's move on to the next question. How many organizations and companies have you made presentations to?"

"Um, well, actually none" the young man admitted, a little dejectedly.

"Oh dear", said the Consultant, quietly. "You really need to get out and speak to groups of people and tell them about your service.

"I don't just mean talk to them individually. I mean arranged presentations where you pass on information. Presentations where you tell the group what you are about and let them know how invaluable you could be to their businesses.

"You could start with the organizations you already belong to. The Chamber of Commerce and Rotary Clubs are always looking for guest speakers."

"Oh no!" Gasped the young man, "Not public speaking! I would do anything for success, but I won't do that!"

"That is a rather illogical statement" said the Consultant. "You know, it's not as bad as you may think and, with a little guidance from the right person, you can overcome your glossophobia— your fear of public speaking; a fear that you share with millions of other people.

"There is a saying that more people would rather lie in the casket than deliver the eulogy; personally, I find that speaking in public is as easy a skill to learn as any other business technique. In fact, in some ways, it is easier and often more vital.

"Now I will recommend you to an associate of mine who can coach you through the steps to take you from a proverbial Jell-O to a confident speaker. Are you willing to take on the challenge?"

"Well... yes, I suppose so. Who do I need to meet?"

"Here is her card. I will phone her and tell her to expect a call from you. In addition I will also be happy to meet with you again in a couple of weeks and we can talk about your progress and also go over your anti-boredom or elevator speech. Thanks for coming today."

The young entrepreneur got up to leave. "Thank you for your help and guidance. I promise I will act on your advice, and I know you will see some changes at our next meeting."

Reflecting on his openly expressed fear, the young man knew, of course, that he could not avoid having to make public presentations. In today's business world there is no place for timidness and he knew that all the successful people use public presentations effectively.

But how was he to overcome this fear? How could he ever hope to make anxiety free, confident presentations? This is what he needed to know and he was sure that his appointment with the Presentations Coach would help him find his Holy Grail.

Abandon All Fear, Ye Who Enter Here

The sign over the Coach's office door was certainly encouraging as he entered a few days later. However, the young man did not feel that he could completely abandon *all* fear, but he was open to listening and being coached.

"I know what you are thinking", said the smiling coach as she entered the room. "I tell people that they only need two things to become effective public speakers. Firstly, they must have the nerve to get up and do it for the first time and, secondly, they need to practise. I can show you how to overcome your fears, and how to put together an effective presentation; after that, it is all up to you."

"My business mentor told me that it is not uncommon to fear public speaking. But why do so many people suffer this affliction?" asked the young man.

"Good question", the Coach responded. "I can only assume that it is the perceived threat of humiliation and the fear of appearing foolish in front of others. Children do not have this problem; it seems to develop later in childhood until it becomes a full-blown phobia—glossophobia—in adulthood. Peer pressure and the constant need to look over one's shoulder to avoid being stabbed in the back by over-competitive colleagues and bosses probably has something to do with it too.

"Like so many things," the coach continued, "it is all a matter of attitude. Many people say 'I couldn't do that', when if fact they mean 'I wouldn't do that'. Anybody *can* speak in public, or do any other feared task; they just need to have the right attitude and *choose* to try."

"Well I suppose I have been guilty of having a negative mind set", admitted the young man. "My business mentor suggested that I need to start doing some presentations to groups to increase my market effectiveness, and I must admit the prospect fills me with fear."

The coach pointed to two small signs on her desk. The first one read:

What the mind can conceive and believe, it can achieve.

Napoleon Hill

"Can you conceive and believe that you are capable of making public presentations, and that there *is* a way to learn how to make them effectively?" she asked.

"Of course!"

"Then you can succeed; although it will take some effort on your part." She indicated the second sign which read:

Attitude

\+ *Aptitude*

= <u>*Altitude*</u>

"As the saying goes, if you want to soar with the eagles, don't hang around with the turkeys. Combining your ability to do something with a positive attitude toward getting it done will ensure you fly as high as you want to", encouraged the coach. "Let me show you something." She reached for a piece of paper and a pen.

"Assigning an ascending numeric value to each letter of the alphabet, with A=1 and Z = 26, we can express attitude as:

$$ATTITUDE =$$

$$1+20+20+9+20+21+4+5 = 100\%$$

"Anything less than 100% attitude leads to failure."

The young man realized it was true.

Impressed, he beamed, "OK, I'm convinced. So, now that we have started the math lessons, will you please show me how to add the aptitude to my attitude?"

"I can certainly lead you through techniques to overcome nervousness and deliver effective presentations, but as I mentioned earlier, it will take practice on your part too. Together we can formulate some material to work with and then I will send you out into the big wide world to 'get your feet wet'."

"Forgive me for asking", said the young man, "but will it be a lengthy process?"

The Coach smiled. "Don't worry; I'm used to working with time-challenged people. The basic process can be learned quite quickly."

"OK. So is it just a case of getting up and doing it?"

"Oh no!" exclaimed the Coach. "No indeed. You may have heard the expression from the property world, 'Location, Location, Location'. Well, for presentations it is a case of Preparation, Preparation, Preparation.

"You need to prepare your material, prepare your venue, and prepare yourself. As another old saying goes: 'failing to plan is planning to fail'.

"There is a great deal of work to do before you open your mouth to speak in public. By following the techniques that I give you, you will be able to overcome your fear easily and make powerful, effective presentations.

"I will show you seven steps to avoid falling flat on your face when making your presentation. Mastering these steps will have you soaring with those presentational eagles and wondering what all the fuss was about."

The young man was intrigued. Being an organized and methodical individual, he liked the idea of seven steps. "So, there are only seven steps that I need to know to become an effective public speaker?"

"Well", responded the coach, "strictly speaking there are lots of aspects you need to consider, but I have collated them into seven easy-to-remember groups. Together, we can build up your confidence step by step. Are you ready to start?"

"OK. I'm ready", he answered, slightly hesitantly.

Step Number 1

Research

- *Your material*
- *The audience*
- *The venue*

"Well, when it comes to my business, I know my stuff", said the young man, confidently.

"Good. So you know that when you present your case to an audience you are judged to be the expert and so, it is imperative your material is researched thoroughly. You can be sure there will be at least one know-it-all in the audience who will challenge your facts or, at least, ask some very testing questions.

"You must study all the relevant trade journals and publications, and you must be able to substantiate all your facts and figures. You must also know how your business differs from that of your competitors.

"You can never have too much information but you will not present everything. You need to concentrate on key points—the need-to-knows rather than the nice-to-knows of your subject.

"Store the rest of the information in your memory ready to bring out as answers to questions."

"Hmm, I can see why you give so much importance to research. I would hate to be caught out by a testing question."

"Don't get hung up on that aspect", reassured the coach, "as I will give you some techniques for dealing with questions later. For now, just concentrate on really getting to know your material, and making sure you have as much research to back your statements as possible."

"OK. But what do you mean about researching the audience and the venue?"

"Let's think about the audience first. They are your primary concern because, after all, the presentation should be all about them and not about you."

The young entrepreneur looked slightly puzzled. "What do you mean by that?" he asked.

"If you focus on *your* needs, such as *your* need to sell your product, then your audience will be unreceptive and you will not achieve your goal. Every member of every audience tunes into the same radio station: WII—FM, or What's In It—For Me?

"They need to know that there is a benefit to them in listening to the presentation. There must be some demonstrated value to them, and it is your job to demonstrate that value to them in your talk."

"So, how do I do that?"

"Firstly, you must research your audience. Find out everything you can about them. What are their demographics? What are their ages, their backgrounds, their levels of experience, even their genders and ethnic origins if this is relevant. Ask yourself, 'What do they know about this subject or about this product already? What do they expect from this presentation: a demonstration; factual information; entertainment; product benefits; freebies?'

"You should liaise with the event organizer beforehand and obtain all these details. You may even want to interview some of the attendees in order to have a clear picture of exactly who you will be addressing. Forewarned, as they say, is forearmed.

"Another thing that you could and should do is to personalize your presentation with specific names, topics and events which the audience can relate to. You can use your own personal stories but you modify them to make them appear completely fresh for *this* audience."

"Wow!" said the young man. "I am beginning to see what a daunting task I have let myself in for."

"I know it seems like a lot to consider, and there is actually lots more too, but don't panic, I assure you that eventually, you will really look forward to the presentation. Once we get through all of this, you will be raring to go! Now, let us consider the venue.

"There are so many aspects of the venue that you need to consider that I formulated a checklist some time ago and find it invaluable. It covers things like the actual speaking location, the time of day, the size and layout of the room, the lighting, amplification and equipment usage. I will give you a copy; please feel free to use it, there is no copyright. My only request is that if you think of anything else to add to the list, just let me have a copy of the revised list."

"Thank you", said the young man. "I'm sure that will be most useful".

"Now", went on the Coach, "we have covered the research side. We have looked at the material, the audience and the venue. I think that is enough information for one day. Tomorrow, we will go onto step two and start looking at how to put this masterful presentation together.

"Your homework tonight will be to make a draft plan of attack. I would like you to put together an outline of the presentation including the main highlights and key points—the need to knows.

"Remember that the brain can only absorb what the backside can endure, so as listeners start to fidget in their chairs their attention span will reduce. Consequently, you need to get your message across in an interesting, clear, logical and brief way. When you bring me the outline we will go over it together."

"I see that I have a lot of work to do. I will bring the outline with me tomorrow and I look forward to progressing on this exciting adventure."

With that the young man left the Coach's office. He was drained but exhilarated at the same time. That night he worked hard to prepare his draft outline.

Step Number 2

Planning the Presentation

- *Content*
  - *Introduction/Opening*
  - *Main Body*
  - *Conclusion/Summary*
- *Delivery Style*
- *Equipment*

The next day at ten o'clock sharp, the young man was back at the Coach's office, presentation draft in hand.

Introduction:
- *Outline of the problematic situation*
- *Offer of a solution*
- *Indication of benefits*

Main Body:
- *Current Situation*
  - *Time*
  - *Money*
  - *Effort*
  - *Paperwork*
- *Possible Solution*
  - *Ease of use – how to*
  - *What can be done*
  - *The when, where and why of it*
- *Benefits*
  - *Efficiency of operation*
  - *Cost savings*
  - *Improvements to morale of users*
  - *Customer satisfaction*

Conclusion
- *Summary – from problem to solution*
- *Request for action*
- *Contact details*

"That really looks good", the Coach beamed. "It certainly gives us something to work with. We may need to flesh it out but as we already know, you are the expert delivering this message so the content will be well known to you anyway.

"For a successful delivery, all you need is a set of notes, probably small cards or a single sheet of paper which you can place on the lectern or table. These notes will only contain key words or phrases, just enough to keep you on track."

"Won't I require pages of typed notes in case I forget things?" asked the young man.

"Not only are pages of notes not required, they have no place in the presentation arena. It is important to remember you are giving a verbal presentation not reading a story."

"Point taken."

"Now, let's take a close look at your draft.

"In any good presentation there are three main elements: content, delivery style and equipment. Let's start with the content

"The first thing to consider is the introduction. Here you need to grab the audience's attention. Your opening words are like the opening sentences of a book. If the first couple of paragraphs aren't interesting, then you will probably put the book down and look for something else to read. The same goes for presentations.

"However, if the opening sentences grab your attention and create real interest, then you will want to read on and, probably, not put the book down until it is finished.

"Your presentation should be the same. You must create that initial interest, something to spark the audience's imagination immediately, and they will want to keep listening rather than drift off into their own little world or, worse still, fall asleep."

"So how do I create that interest?" asked the young man.

"There are several ways. You can start off with a quotation, a startling fact or a statistic, or maybe some humorous story.

"However—and this is a big thing to remember—whatever you choose must be relevant to your message, and must flow smoothly into the main body of your presentation.

"I think that speakers who start off their presentation with the latest internet joke, which usually bears no connection to the rest of their presentation, are just plain lazy. They should be dragged away from the microphone; most audiences seem to agree, they often lose interest in such a presentation early on.

"Remember, you are trying to 'hook' your audience and get their undivided attention before you can reel them in with your message. Now, let's look at the main body of your presentation."

Something about the Coach's approach struck the young entrepreneur. He asked contemplatively, "I notice you keep referring to it as a presentation rather than a speech. Is there a reason for that?"

"That's a great observation", responded the Coach. "I believe this particular medium of getting a message across to a group of people should be a presentation and not a speech.

"A speech is something that is often delivered from behind a lectern by someone who is probably following a script that cannot be deviated from. Often a speech is written by a speech writer so that the message is delivered in a very precise, carefully considered manner. The speaker may deliver it in an interesting way, but it is still a speech and not a presentation.

"A presentation, on the other hand, is delivered less formally, often with more passion or conviction and some humour. A good presenter connects with the audience, making the whole experience enjoyable for speaker and listener alike."

The young man smiled. "I have certainly suffered through some of those bad speeches and know what you mean about the enjoyment factor."

"So, now that you have hooked the audience with your interesting opening, it is time to start delivering your message in a logical, informative and interesting way so that they remain interested.

"I see from your draft that you have three main points to your presentation: the current situation, a possible solution (your product or service), and the benefits. That is good.

"The main body of a presentation should contain only about three or four key points, occasionally five if they are not too long. A maximum of five points is about all that a typical audience can absorb and retain, and you need your audience to absorb and retain your message until you get to the call for action.

"Remember, just give the audience the main facts. Back these up with complementary information, statistics and data, and leave them wanting more. If they are interested, they will come to you to ask for details, or talk contracts."

"I have heard the phrase 'Less Is More' but I have never thought of it in terms of a presentation before", commented the young man

"This is a classic example of that saying", responded the Coach. "Now, let's work on the ending of your presentation.

"Just as the opening was like a hook to get their attention, the ending or conclusion must leave them with a memory.

"It has been said that the most important minute in a ten minute presentation is the eleventh minute. That's when the audience goes away remembering what they heard.

"Consequently, it is important to summarize what you have said and to deliver that summary in a memorable and interesting way."

"So how do I do that?"

"There are several techniques. You can recite the key points chronologically, and remind the audience of the consequences and benefits that you outlined.

"However, while sufficient, that method tends to lack the passion you have tried to show in the rest of your presentation. Rather than recite the key points you could come up with a short, simple phrase which captures the essence of your message and which lingers in the mind of the audience.

"You could also use a quotation or short story, but make sure it is one that ties into your opening, thereby tying the whole thing up nicely, and helping your audience to remember the entire message."

"Wow! All that and we're still only looking at the content."

"I know it seems like a lot to remember, but as you practise, the whole thing will just seem like what brushing your teeth is to getting up in the morning; you just do it—almost without thinking.

Let's go over the content once again."

Content

The Opening = The Hook that grabs the audience's attention.

The Main Body = The Facts or key points—backed up with data, stories, statistics and evidence.

The Conclusion = The Memory; a summary and/or call for action.

"Well, I think I have that down pat now. What do we have to do next?" asked the young man, eager to move on.

"Remember", laughed the Coach, "the presentation is you. How you deliver it reflects your style, your values, your beliefs and your personality. If you want an audience to make use of your product or service, you must make them want to do business with you.

"Your presentation may be all about what you are offering, but your delivery is all about selling you. Let's face it, if you were buying a $10,000 piece of jewellery, you would not expect it to be wrapped up in a plastic supermarket bag, now would you?"

"Fair comment. So, now that we have the message in the best possible format, I suppose all that remains is to produce some PowerPoint slides and then practice the delivery."

The Coach groaned and raised her eyes to the ceiling. "Not so fast. We will come to the delivery style and your platform performance soon enough. But let me emphasize as strongly as I can, there is so much more to an effective presentation than a bunch of PowerPoint slides."

The Coach gestured towards another sign on her desk.

Be a Lenient Judge of Presentations. Do Not Sentence Your Audience to Death by PowerPoint.

"Now don't get me wrong," she continued, "projected slides have their place, particularly in a training session where certain information needs to be conveyed in a visual manner to the learners. But you are not giving a training presentation."

"Oh! Then I guess you have an alternative?"

"Indeed I have", beamed the Coach. "Step three."

Step Number 3

Use Visual Aids Sparingly

"Presentations", she continued, "fall into three types: high-tech, low-tech and no-tech. These are fairly self-explanatory terms, but let's look at them a bit more closely.

"A high-tech presentation features projected slides controlled by a laptop computer using a liquid crystal display (LCD) projector onto a large screen.

"The slides often feature many bright colours, graphics, and words. These are often flying in or fading out from all directions and accompanied by a variety of sounds.

"While these presentations can be great, they can also be deadly boring. We have all been at presentations where the presenter walks between the projector and screen causing a large shadow over the image, or even worse, when he or she reads off the screen as if the audience is too stupid to be able to read. As you can tell, I am not a great fan of these gimmicky presentations.

"The whole idea started off well intentioned enough, but then people began exploiting the system and overusing it. It is best to remember that just because PowerPoint *can* do something, there is no reason why it *must* do that something."

"But nearly everyone uses PowerPoint in their presentations these days."

"Exactly," the Coach agreed, "but you don't want to be 'everyone'. You are unique and you want your presentation to be unique too. So let's concentrate on making it appear unique."

"So, do you mean that I shouldn't use any slides?"

"If you think that a limited number of slides will add to your verbal presentation and not detract from it or, worse, start to overpower it, then by all means create some.

"Visual aids should be exactly that: aids. They should enhance your presentation not take it over. They should be used sparingly and should never become your set of presentation notes.

"As I mentioned before, I don't like it when presenters project slide after slide, each one filled with rows and rows of words, and maybe the odd picture, and then proceed to read all of those words. This causes the presenter to lose credibility and become ineffective.

"Do not insult your audience by just reading off the slides. Instead, provide them with a handout to take away. They will thank you for your consideration and you will save yourself the effort of talking too much.

"If you must use projected slides, then make use of pictures, graphs and diagrams, and just a few key words, all of which you can talk about in your presentation. That way you will remain the focus of your audience."

"Wow." The young man looked a little pensive, but as he thought about all the Coach had told him, he brightened. "You are right. I do need to differentiate myself from everyone else. That is how I will be successful."

The Coach smiled. "Now, if you do decide to use any equipment at all, you must know exactly how it works. You must also have a contingency plan for when it breaks down. That's when, not if. But, as I said before, high-tech is not the only option. There are the low-tech and no-tech options too.

"Low-tech doesn't mean low quality, it just means a little less reliance on technical gadgetry. For example, you could make use of a flip chart to write or draw on. However, if you are presenting in a large room that may not work too well, as people at the back of the room will not be able to see the chart. Remember what I said about how you must research your venue? Well, that includes checking to see how well any visuals can be seen by the entire audience.

"If appropriate, you may consider showing a model or having a large picture set on an easel. You could give out handouts, although it is prudent to wait until after your presentation to give them out. You do not want your audience reading while you are talking."

"OK, I'm beginning to get the picture. So, no-tech means that I have to rely on my voice to capture and keep the audience's attention, without resorting to any visual gimmicks."

"Got it in one! And another point to remember on using visual aids: you may have a picture or an object that you want everyone to see. By all means show it to the audience and then put it out of the audience's line of sight, unless you need to make constant reference to it.

"Why out of sight?"

"For two reasons", the Coach asserted. "If it is constantly in sight the audience might keep looking at it and not at you. They might lose their concentration and your message may become lost.

"You may also be tempted to keep looking at it yourself, and in this manner you can't be making appropriate eye contact with your audience. They might think that you were not talking directly to them, and maybe they will translate that into the idea that you are not really interested in them. Again, your message could get lost.

"However," she continued, "if you want the audience to see the object more closely, you could always invite them to come to you after the presentation. That way, you keep control of the presentation *and* you get to chat to them more intimately after the presentation.

"Similarly, if you want to give out any handouts, do so at the end of the presentation. That eliminates the inevitable change in concentration as the audience stops listening and starts reading when you pass out the information during the presentation."

The young entrepreneur felt another stab of pensiveness. "You mentioned eye contact there. Now I know that it is essential to look at your audience when you are talking to them but it sounds as though you have something more controlling in mind. Eye contact is something that worries me."

"Don't worry" the Coach laughed, "you are already well on your way to making effective presentations.

"OK, so now we are moving onto the next element of your learning: step four."

Step Number 4

Presentational Techniques

- *Voice*
- *Eye Contact*
- *Gestures and Body Language*

"Here we are going to focus on the actual presentational style and tips on delivering your message. There are three things to think about here, the voice, eye contact, gestures and body language, and I want to deal with each of them in that order. Look at this."

She handed him a large white card.

The voice of the intellect is a soft one, but it does not rest till it has gained a hearing.

Sigmund Freud

"A very profound statement which really reflects what giving a presentation is all about", said the Coach. "You see, most people are afraid of standing up and speaking in front of a group. They don't want to make a noise and as a consequence, they don't make themselves clear. Usually, that is because they don't know how to.

"Remember, your voice is your main tool and, like any craftsman, you need to use it to its maximum capacity."

The Coach went over to a white board on the wall and wrote four Ps in a vertical line. "Correct use of the voice", she said, "involves four P-elements. The first is Power." She wrote the word Power on the board.

"In a presentation you need to speak only slightly louder than you would in a normal conversation. But just as in normal conversation you must vary the volume and use voice modulation. Sometimes you speak quite loudly and sometimes, quite softly. By varying the volume of your voice, you can make more impact with your words; also this variation makes your voice more interesting to listen to.

"Try not to tail off your voice toward the end of a sentence or a point. When we talk, especially in North America, we tend to drop our voices at the end of a sentence. Sometimes this means that the audience won't hear every word that you say. When you are checking out the venue, make sure you test the acoustics and ensure your voice carries to everyone, no matter where they are sitting."

"What if I am using a microphone?"

"The same answer applies. However, you must test out the equipment before the audience arrives just so that you are comfortable with it.

"The second P is for Pace", she said as she turned and wrote the word on the board. "The Pace, or the speed that you speak at, should be slightly slower than in a normal one-on-one conversation. Make sure your diction is very clear.

"People need to be able to hear every word. This is particularly important if you have an accent—and everyone in the world has an accent! As with your volume, vary the speed of your delivery. Sometimes speaking more quickly and sometimes more slowly, depending on the emphasis you are making."

"I suppose", mused the young man, "that these vocal variations are designed to keep the audience's attention."

"Absolutely. Good observation. And this is even more important with the third P which stands for Pitch, sometimes known as tone." This third P was written on the board.

"There is nothing worse than hearing a monotone speaker. It guarantees to send the listener to sleep. You need to vary your voice tone—sometimes high and sometimes low. Use intonation and voice inflections to make the words more interesting.

"Most of all, show enthusiasm in your voice. Speaking in public is a bit like singing without the music."

"I never thought of it that way before, but that is a good analogy.

"So let me try to guess the last P. Projection?"

"Good guess but wrong."

"Phonetic?"

"Good guess again, but still wrong."

"OK, I give in."

"It is something that many new speakers are afraid to use, but something that can be one of the most effective and powerful tools when used correctly." Onto the board she wrote the word Pause.

"Wow!" he said, surprised. "So, how can a pause be so powerful?"

The Coach returned to her seat and sat down. "By inserting a pause into your message you achieve one of two things.

"Firstly, it allows the brains of the listeners to catch up with their ears. People hear things very rapidly but their brains sometimes need a little longer to process what has been heard and create a response.

"Secondly, you can use the pause for effect. Leaving a gap between words can inject a note of eager anticipation for what is coming next."

"But doesn't it sound like I have forgotten something?"

"Absolutely not", explained the Coach. "Pauses are a deliberate part of speech. I am not talking about the sort of gaps usually filled by 'ers', 'ums', 'ahs', 'and ands', 'you knows' or any of those other little fillers.

"Those are just nervous noises translated into something like, 'Just hang on a moment listener, I am going to say something more significant very soon. Honestly!' Those sorts of noises also often show that the presenter has not practised."

"Ah, ha! I had not thought of it like that before. It is surprising how many people do use those type of fillers."

"Yes indeed. In a normal conversation you can get away with it to a certain extent, because it is expected that you are constantly thinking and speaking at the same time.

"However, in a prepared presentation there is no room for such fillers. The more the audience hears of them, the more likely they are to lose concentration in what you are saying as they start to concentrate on the 'ums' and 'ers'."

The Voice

P—power
P—pace
P—pitch
P—pause

"Okay, now, let's talk about the eye contact; we touched on it earlier."

"I guess there is more to it than simply looking at the audience."

"Yes and no", said the Coach as she took a torn-off calendar motto from her pin board.

You cannot depend on your eyes when your imagination is out of focus

Mark Twain

"As a presenter," she said, "you have to imagine that you are going to do well. You have to imagine the audience seeing everything through your eyes. In order to do that, you have to be able to see through *theirs*.

"Every person in your audience needs to know that you are talking directly to them—as an individual. To accomplish this, you must make eye contact with every one of them.

"Now, with a large audience it is difficult to be that intimate. However, you still need to direct your attention towards every section of the audience. You need to pick out individuals with your eyes and deliver part of your message directly to that person. People sitting close to that person will also think that you are speaking directly to them.

"Many speakers tend to have a bias toward one side of the room or the other. That is, they predominantly look towards people in, say, the right side of the room. When speakers do this, people in the centre and on the left side of the room feel the speaker is not speaking to them. Why? Because there is no eye contact. The result is that those people switch off; they stop listening."

The young man thought about what the Coach had just said. "Yes, now that you mention it, I have seen examples of that. I've even been in the audience when that happened and it's true, I did feel excluded. I had no idea that the eyes could hold such power".

"I said earlier that your voice was your main tool. Well, your eyes come a pretty close second. Aided and abetted by your facial expressions and body language; they help you control your delivery and connect with your audience.

"There are, however, a couple of tips I'd like to give you to help you maintain effective eye contact. Firstly, as you look around the room at different individuals, keep the sweep of your gaze random. Don't be like a lighthouse, moving your head from one side of the room to the other and back again in a steady pattern. If you use a random approach, everyone will feel included and you will be observed as a caring and effective presenter.

"Secondly, don't look at any one individual too long. Keep the randomness uniform. In other words, be consistent with your random search pattern; look at every individual in turn in every section of the audience; but keep your gaze on a single person to just a few seconds and then move your gaze on.

"If you focus too long on an individual, such as a nice friendly face in the front row, two things will happen. One, that individual may feel embarrassed and go red as they wonder why you are speaking to them so much. Two, the rest of the audience will stop listening because they will think that you are not talking to them. They, too, may wonder why you are giving one person so much attention."

"Once again, wow!" sighed the young man. "I am beginning to see why so many people dread doing presentations, there just seems to be so much to remember."

"Now, don't worry, you are doing great."

"You also mentioned facial expressions and body language. Tell me about those aspects please."

"Certainly", smiled the Coach, impressed with the young man's dedication.

"The face is a wonderful conveyor of emotions. It can show happiness, sadness, excitement, frustration and every other possible feeling.

"Most people use facial expressions naturally; so, in a presentation, all you have to do is make those expressions slightly more exaggerated so that the audience can see them. The most important facial expression you can show is a smile! If you look as though you are enjoying the presentation, then the audience will too.

"Gestures and body language can also be used effectively to add emphasis to the words being spoken. Now I am not talking about wild, flailing arms resembling an out of control windmill. Nor am I talking about the repetitive raising and lowering of an arm and looking like a Las Vegas slot machine. I am talking about natural gestures that we use all the time when we speak."

Cautiously, the young entrepreneur asked, "Like most people, I wouldn't know what to do with my hands in a presentation. What would be your advice?"

"Good question", responded the Coach. "First, you should decide on a comfortable 'rest' position.

"There is nothing wrong with arms resting gently down by your sides, but you may want to raise them a little and bend the elbows so that your hands can rest open in front of your tummy. From there they can move naturally into meaningful gestures. This may sound awkward, but with a little practice it will seem quite natural.

"Keep your hands out of your pockets and never play with coins or keys. Also, avoid holding anything in your hands, just in case you drop it. Of course, if you are demonstrating something, that is quite different. But whatever you do hold, make sure it has a purpose and always put it down as soon as its usefulness is over.

"As you practise your presentation you can start adding deliberate gestures for effect. But any gestures must look natural and not over rehearsed. And that brings us nicely to step number five."

Step Number 5

Practise, practise, practise

"Practice is what will turn you from a mediocre presenter to an effective one. Practice is really essential if you are to become completely familiar with your material, and with the way you will present it.

"Also, practice will help you develop your stage presence".

"What is that?" the young man asked pensively.

"Having a good stage presence is when you project that you are quite comfortable in front of an audience—even if you are nervous! It is how you take command of the stage, and how you are able to move around your space comfortably, meaningfully, using body language that says 'I am a great, but human, presenter'."

"Human?"

"Yes indeed. Even though you are the expert when you are presenting, you do not want to come across as either a boffin or a buffoon.

"Boffins are stuffy, know-it-all, unapproachable people who behave as if they have recently undergone charisma bypass surgery.

"Buffoons, on the other hand, look like they are more used to playing a supporting role in a shabby B-movie. It is easy to see that they missed the Practise, practise, practise.

"Of course it is alright to make occasional mistakes, and it is perfectly okay to interact with the audience, especially when you get to feeling more confident. I actually suggest such interaction in step six."

Step Number 6

Interact with the audience and enjoy the experience.

"Once again," she continued, "you maintain the connection with your whole audience. You may ask questions to elicit a response from your audience, or you may invite and respond to questions from them.

"Be careful on that last point though. Answering questions during a presentation can make you go over time and may throw you off track. It may be better to suggest that questions wait until the end of the presentation."

"Is there a special way to deal with questions?" he asked

"Yes, particularly when you are speaking to a large audience.

"The first thing to remember is when the audience is asking questions, it shows that they are interested in what you are saying, and that is a good thing.

"Let's go over to the board and summarize the techniques I recommend on answering questions", she said as she moved to the white board. She wrote the following list:

Dealing With Questions

T—*thank & acknowledge*
R—*rephrase*
A—*answer*
C—*clarify*
Y—*you move on*

"Tracy can help you in dealing with questions," said the Coach with a smile on her face.

"The first thing you should do is acknowledge the questioner in some way; thank them for the question or perhaps say, 'Good Question'. You want to make them feel that their involvement is valued. After all, they are probably asking a question that many others in the audience would also like answered but are afraid to speak up and ask. (Remember all those people who hate to speak in public?)

"You may need to rephrase or repeat the question. This serves two purposes: it clarifies the question in your own mind, and it lets the rest of the audience know what was asked. If the audience is small and everyone can hear the question clearly, then you may skip this repeating, but it may still be worth rephrasing the question just so you can be certain what is being asked.

"Of course, you must answer each question in a clear, comprehensive manner, in a way that is understood by the questioner. Once you have answered the question, you may want to check with the questioner by saying 'Is that OK?' or 'Does that make it clear?' Then you move on to the next question or wrap things up."

The young entrepreneur looked pensive again. "What about the awkward question or the one I don't know the answer to?" he asked.

"As the 'expert' you should really know the answer to any question that is asked of you. That is, after all, part of your research to anticipate all questions.

"However, if there is something that really stumps you, apologize that you do not have an immediate answer, but promise to find the answer by a certain time, and arrange to meet, phone or email the person who asked the question with the information. And that would go for anyone else who needed to know the answer to that question. Make sure that they provide you with their contact details after the presentation.

"If someone is trying to trip you up with a question which has no relevance to your presentation, respond by saying that it is an interesting question and one which would require a fuller answer than your current time allows. Then suggest that you meet one-on-one with the person afterwards to discuss the matter before turning to take the next question. This politely takes the focus off the person asking the question and lets you keep control."

"You know, you have given me so much information that I am not sure that I am going to be able to remember it all."

"Do you drive?" asked the Coach.

"Yes, of course", he responded, somewhat taken aback.

"Remember when you first learned to drive. You probably got in the car thinking that all you had to do was hold on to the steering wheel, press a couple of pedals and everything would be OK.

"You were in a state of what is called 'Unconscious Incompetence'. You didn't know what you didn't know. Then someone showed you exactly what to do, how to start, how to move, what to look for, when to brake, when to turn, how to steer, how to stop. How to be safe.

"Suddenly you realized that there was so much to learn. You knew what you didn't know and entered the state of 'Conscious Incompetence'.

"Then as lessons progressed you became more confident and knew what to do, when, and why. You moved into the state of 'Conscious Competence'. You knew what to do but you still had to think about each part of the process.

"Now, have you ever been driving along and suddenly thought, 'How did I get here? I can't remember going through that usual stop sign. I can't remember turning at the store as I normally do.' Well, my friend, you have moved to the fourth and final stage of the learning process called 'Unconscious Competence'.

"You do things without even thinking about them. Everything you learn to do in life follows this pattern."

"I really appreciate that", he said. Thank you. That is a great analogy and I'm beginning to feel better already."

"Good. Right now you are rather scared because it seems like there is so much to take in and learn. But it will get easier, and probably faster than you think. The next stage of the process is learning by doing. As with all learning, the more you practise, the better you will become. And I will be here to help you with the practising."

"Today, we will finish off with some tips to help you control the initial nervousness you may well experience. Step number seven is – breathe!"

Step Number 7

Breathe!

"Ha, ha!" laughed the young entrepreneur. "Now, that seems very obvious but I suppose you have some deeper interpretation."

"Yes and no", responded the Coach. "It is surprising how many inexperienced speakers apparently forget to breathe, or at least to breathe properly, during their presentations. It is almost like they are racing to get through the presentation and get off the platform—and of course if they haven't practised, they are probably trying to do just that.

"Now that you have gone through my program, this will not happen to you.

"Here are some tips for you to remember to help you control your nervousness before and during your presentation." Again she moved to write on the white board.

Tips for Controlling Nervousness

- *Know your material*
- *Practise*
- *Focus and pre-visualization*
- *Positive mental attitude*
- *Stretching exercises*
- *Breathing exercises*
- *Enjoy*

"It goes almost without saying", she said, "that you will know your material thoroughly and that you will have practised the delivery several times in advance.

"In addition you will get yourself in the right frame of mind by focusing on a successful outcome. Many successful athletes and performers, do something called pre-visualization before an event. They 'see' the whole performance or race in their mind's eye. They 'see' everything, step by step, section by section. They visualize every little detail being completed perfectly and of course they see themselves winning the race and receiving rapturous applause.

"See it, believe it, do it—make it happen."

"Yes", responded the young man, "I have seen athletes warming up at the start of a race staring down the track before they get on their blocks. That must be what they are doing. Pre-visualizing the race before it happens."

"Exactly. It is essential to maintain a positive mental attitude. If you think you will do well, or if you think you will do badly—you are right!

"If you think positively, then you will be better equipped to overcome any problems you encounter and complete your task successfully. If and when things do go wrong, just think of it as serendipity.

"Remember, both stage fright and stage excitement use the same amount of adrenaline—it is just your perception of the situation that makes it either hard work or a pleasure."

"I had not thought of it that way before", said the young man. "Many people talk about stage fright but I guess you are saying think positively and use positive language."

"You are catching on quickly", the Coach enthused.

"A couple more tips for you. Just before you get on the platform to speak, do some stretching exercises. These can be done either sitting or standing.

"Stretch your arms and legs as far as you can; make yourself feel taller. If possible, move your head in a circular pattern—down, up and round one way, then the other. Clench your fist then stretch your fingers out wide a couple of times.

"Breathe deeply two or three times. Breathe from right down in your diaphragm. Breathe in through your nose and exhale slowly through your mouth. Don't hyperventilate. This exercise is best done with your eyes closed.

"If you follow these tips, you will feel more relaxed and ready to do your best. Even when you are on stage, if you feel a little tension rising, do a surreptitious stretch or take a deep breath. The audience will never notice and you will instantly feel better.

"My final tip is this: when you are introduced, bound onto the stage or platform with enthusiasm and a smile. Be Dynamic! Show your audience you are thrilled to be there with them."

"I never thought I would say this," hesitated the young man, "but I think that I can actually do this."

"Of course you can", laughed the Coach. "And in the next couple of meetings I will help you to practise prior to your first presentation. But for now, let's take some time to summarize the seven steps.

7 Steps to Presentational Success

1. *Research (material; audience; venue)*
2. *Plan (outline; delivery style; equipment)*
3. *Use visual aids sparingly*
4. *Presentational techniques (voice; eye contact; gestures and body language)*
5. *Practise, practise, practise*
6. *Interact with the audience*
7. *Breathe!*

"Thank you very much for all your help and guidance," enthused the young entrepreneur. "I have learned so much already and I look forward to practising my technique under your watchful eye. No more Mr. Shy Guy for me from now on."

The Coach smiled warmly at him. "I have enjoyed working with you too. Come back in a couple of days and we will make a start on turning the theory into practice. Good-bye for now."

The young man went home, exhausted but excited. He continued to work with the Coach until he felt really comfortable with his presentation material and delivery style.

Together, they arranged for him to address a meeting of the local Chamber of Commerce but before the big event he knew that he had another visit to make.

He returned to the office of his Business Mentor to report on his progress.

"I'm glad you took my advice", beamed the Mentor as he heard what the young man had to say. "Now you are ready to take your business to greater heights. Now you can expect success."

"Yes, although I still have some butterflies in my stomach, I know that that is excitement not fear. As my Coach said, using her seven steps, those butterflies will soon start to fly in formation.

"What is even more amazing to me is the fact that not only did I think I couldn't make presentations, but I thought I was far too busy to ever take time to learn. Now I realize I can't afford not to make the time to learn; not if I want my business to be successful."

"Ha, ha!" laughed the Mentor. "You are absolutely right. I can see you are a changed man from when we first met. You no longer dread the thought of making a presentation, and now you understand that learning how to do this properly is a real time saver, as well as a business saver."

"Absolutely right. I no longer suffer from glossophobia. In fact," said the young entrepreneur firmly, "now I can honestly say, I would do anything for success and I certainly will do that!"

Venue Check List

Location:
- *Directions/Map*
- *Time needed to get to venue*
- *Transportation arrangements*
- *Arrangements for access to venue prior to / on the day*
- *Contact name/phone number*

Timing:
- *Start/finish times*
- *Meal arrangements*
- *Refreshment breaks*
- *Ice breakers/warm-ups*
- *Clock locations*

Heating & Lighting:
- *Access to controls*
- *Spotlights/operation*
- *Room lights/operation*
- *Briefing details to lighting operators*

Size & Layout of Room:
- *Location of stage/lectern*
- *Table layout*

- *Props table*
- *Access to audience*

Sound:
- *Sound control/operator*
- *Briefing details for sound control operator*
- *Lectern mic*
- *Handheld mic*
- *Lapel/body mic*
- *Audience mic*
- *Sound check arrangements*

Equipment:
- *OHP/Projector/Screen*
- *Flip Chart/paper/pens*
- *Table for handouts etc*
- *Handouts*
- *Books*
- *Pointer pen*

Acknowledgements

As we progress life's journey, we encounter many people along the way. Almost without realizing, we are influenced by every person we meet and everything we become involved in. How we react to those influences affects our very being.

I have been extremely fortunate to belong to a number of wonderful organizations which, through my active involvement, have helped me gain the knowledge and experience necessary to write this book. Consequently, I would like to acknowledge, in particular the following organizations:
Rotaract (www.rotaract.com; www.rotary.org/programs/rotaract); Junior Chamber International (www.jci.cc); Toastmasters International (www.toastmasters.org); Canadian Association of Professional Speakers (www.canadianspeakers.org); and everyone I have met through those groups.

I would also like to acknowledge Kenneth Blanchard PhD and Spencer Johnson MD, the authors of "The One Minute Manager". Their book greatly influenced me in the design of this book as well as helping me in my business life.

Finally, I would like to acknowledge the help received from Terry-Lynn Stone in editing this book and Michelle Brown (www.mbart.ca) for designing the cover. Also, to my wife Sandra and son Simon, for their help, support, encouragement and love.

David Hobson

About the Author

David Hobson is an award winning, international speaker and professional trainer in business communication skills, with a background in commercial and industrial training design and presentation.

He specializes in training and coaching presentation skills, running meetings, trainer training, listening and negotiating skills. In addition, David presents Keynote speeches and Master of Ceremonies services.

Having over 25 years of experience in public speaking, David is a Distinguished Toastmaster with Toastmasters International, a member of the Canadian Association of Professional Speakers and was trained in the English traditional style of Professional Master of Ceremonies.

As a winner of several public speaking contests, and recipient of the 2003 Presentation Mastery Award from CAPS Vancouver, David is able to inspire and entertain audiences with a mixture of motivational, humorous, informative and interactive presentations.

David's favourite book of all time is "The Hobbit" by J. R. R. Tolkien. The tale of a very shy person, who is coerced into going on an adventurous journey through strange and exciting places; meeting with weird and wonderful characters; overcoming a great many obstacles; finding a rich reward; and returning to tell the story; is very reminiscent of David's life too. As a consequence, he chose to have the name Hobbit Communications (www.hobbitcommunications.com), when launching his business enterprise.

ISBN 1412072034